THOUGHTS ON THE ORIGIN OF LIFE AND OTHER ESSAYS

BY

RB RAIKOW

Reading Glass Books 1 888 482 4596
www.readingglassbooks.com
orders@readingglassbooks.com

Contents

"The only issue I have about the depiction of DNA on the front cover is that the shiny veneer makes the bottom turn look 'mixed up'. But then I thought any representation of the famous double helix is bound to be imperfect, after all we still don't understand everything about the hereditary material."
RB Raikow

1. Thoughts on the Origin of Life

Living cells are organized to an astonishing degree, containing structures of specific shapes, ranging in scale from nanometers to micrometers.

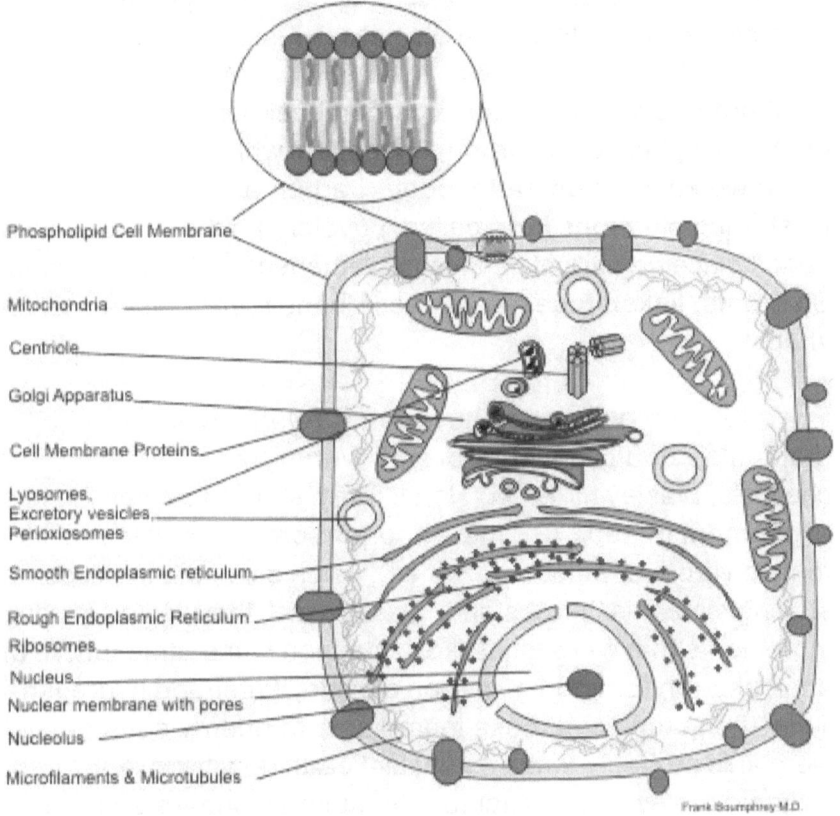

Phospholipid Cell Membrane

Mitochondria

Centriole

Golgi Apparatus

Cell Membrane Proteins

Lyosomes,
Excretory vesicles,
Perioxiosomes

Smooth Endoplasmic reticulum

Rough Endoplasmic Reticulum

Ribosomes

Nucleus

Nuclear membrane with pores

Nucleolus

Microfilaments & Microtubules

Frank Boumphrey M.D.

Diagram of a eukaryotic cell. (Wikipedia)

Here I try to examine whether we have sufficient evidence that physical forces and molecular chemistry, coupled with Darwinian evolution are sufficient to account for this organization.

Radmila Boruvka Raikow

Organic Molecules

Chemistry (1) is classified as inorganic or organic. Inorganic chemistry relies on statistical behavior of countless, small units. Organic chemistry deals with molecules that are relatively large and have defined shapes. These shapes play important roles in their interactions with other molecules.

Most organic molecules have carbon backbones and they are the building blocks of carbon- based life, which is the only kind of life we know of on earth today. Carbon atoms link together by sharing electrons in so called, covalent bonds (2). Covalent bonds place constraints on the movement of the partner atoms. So atoms, linked together by these bonds, are held in certain shapes.

For a long time it was believed that relatively large organic molecules could only be formed within living systems, but chemists, Stanley Miller and Harold Urey, showed by their famous experiment in 1952 (3), that important organic molecules can form in vitro (i.e. without enzymes), when given the appropriate starting materials: Into a sterile, glass vessel, they sealed starting ingredients that were most likely common in the atmosphere of ancient earth (ammonia, carbon dioxide, water and hydrogen). They then subjected these ingredients to energy sources that were also probably found on ancient earth (electricity and heat). Organic molecules crucial for life, such as sugars and amino acids, formed in their experiment.

Proteins

Proteins are linear chains of amino acids, and can be structural (4) or catalytic (i.e. enzymes). Enzymes are essential because

2

they enable organisms to carry on reactions at temperatures and at efficiencies compatible with life. In order words, they reduce the threshold energy needed to start chemical reactions.

Amino acids are so named because each of them has an amino and a carboxyl side group, by which they link together to form proteins. This linkage, called peptide bonding, involves the removal of a water molecule, a process that in today's living cells requires several enzymes as well as three kinds of nucleic acids (discussed below). Until recently some doubted whether amino acids could link up spontaneously to form proteins. However, in 2019, Moran Frenkel- Pinter et al. (5) found that some amino acids will link up without enzymes, under acid, dried- down (i.e. concentrated) conditions. Such conditions could have been found in prebiotic earth.

A wonderful professor at NYU, Douglas Marsland, made a statement in a lecture that has stayed with me for years: He said that "proteins fold into useful shapes, by spontaneous assembly", i.e. proteins fold into characteristic, three-dimensional macromolecules because of their specific amino acid sequences. They assume shapes that stabilize their originally floppy molecules (6) by hydrogen bonding and Van dear Walls forces (7). The important result of protein folding is the formation of pockets or "active sites", in which reagents are oriented in ways that induce them to combine or split. This is how enzymes catalyze metabolic reactions. The sequence of amino acids in proteins, which determines their folding, is formed under the direction of the genetic code, which is written in giant nucleic acid molecules.

Nucleic acids

Biological nucleic acids are called DNA and RNA. Their functionally crucial parts are four nitrogenous bases, Adenine, Guanine, Thymine (or Uracil, 8) and Cytosine. Before linking up into long chains, these nitrogenous bases combine with a phosphate and a sugar (ribose or deoxyribose, 8) to form nucleotides. Nitrogenous bases and sugars were found among the products of Miller/Urey type experiments (see above), but nucleotides, were not. However, in 2016 a report of nucleotide formation in a laboratory was published by Cafferty et al (9). These authors used plausible prebiotic conditions, sugars, inorganic phosphate, and nitrogenous bases similar to present day nitrogenous bases, to produce two nucleotides which then linked, without the help of any enzymes. Interestingly, nucleotides have also been found in meteoroids (10).

DNA chains are the master scripts of the genetic code in all living organisms. (There are viruses whose genetic information is stored in RNA. However, these RNA viruses must rely on the DNA of their hosts to grow and reproduce. Therefore these, as well as all known viruses, are not considered to be really, i.e. independently, alive.) In cells DNA is kept in central areas (in eukaryotic cells within membrane bound nuclei) where they cannot be easily changed by everyday reading. Copies of DNA sequences, in the form of messenger RNA (mRNA), move from these central areas to be read or translated into protein chains.

mRNAs, are fragile long chains and they are usually coupled to transfer proteins in order to ensure they survive the journey from their formation to where in the cell they are translated. Once their sequence is translated into proteins, mRNAs are degraded because they have no stabilizing tertiary structure. The sequence of amino acids in the resulting proteins is dictated by a universal genetic code. (How scientists elucidated the genetic code will be described below.)

Two other RNAs are crucial in all cells' protein synthesis machinery (11). Unlike mRNA, these RNAs acquire tertiary structure or shape by folding. This is possible because RNAs, unlike DNAs, have flexible backbones (12). The folding of transfer and ribosomal RNAs is vital for their role in protein synthesis. Their three dimensional shapes are stabilized by the reactivities and positions of their subunits or nucleotides (This is similar to the way the positions of various amino acids stabilize the folding of proteins).

So, again, we encounter the importance of shape, which is in turn determined by the sequence of subunits in chain-like molecules. This will be the key in my argument but before I continue that, I need to consider other processes and elements essential for life.

All life forms need energy.

Archaea (described below) harness energy from the motion of electrons in inorganic compounds such as sulfur dioxide or use energy released when atoms of small organic molecules are rearranged. Archea (some can be found on earth today) may have been a crucial first step in life's origin. Similar energy release occurs in today's, enzyme-driven Krebs cycle, which given enough starting material is self-replicating. Stubb et all (13) have proposed a prebiotic (enzyme free) cycle that is similar to the Krebs cycle, found in all cells.

Two of the subunits of nucleic acids (adeninosine and guaninosine), can also link up with phosphate side groups by high energy bonds, and become the source of immediately usable energy within all cells (14). ATP (adenosine triphosphate) and GTP (guanosine triphosphate) are crucial cofactors in anabolic reactions within all cells. As in all metabolism, the delivery of this energy, as well as its storage (i.e. the formation of ATP and GTP), is regulated by appropriate protein enzymes, which are in turn specified by the genetic code within nucleic acids.

Concentration of life's ingredients

Cell membranes protect and concentrate the chemical ingredients and reactions necessary for life. The outer cell membrane, or boundary around all cells, as well as the internal membranes found in eukaryotes (see diagram at the beginning of this essay) are made up mostly of phospholipid molecules. When these phospholipid molecules are placed in water, they self- organize into ball like structures called micelles (15). They so organize because of their bipolar, or amphiphilic nature, i.e. their phospho-end has affinity for water, and their hydrocarbon end has affinity for lipids (including their "fellow" hydrocarbons). So, phospholipid molecules in water hide their hydrophobic ends by forming themselves into micelles. Although micelles are small, ball-like structures that have no contents, when they are abundant they form sheets and can become containers, called vesicles. (Vesicles often form at the surface of cells where they transport materials in and out of the cell boundary.)

Interestingly aggregates of phospholipids have been found in outer space, which has led to the intriguing hypothesis (16) that they could have encircled macromolecules formed on earth, after such macromolecules were sprayed upward into the cold upper earth atmosphere. (Such spraying could be have been produced by large meteors.)

Robert Hazen (16 and 17) has proposed a way in which small molecules may be concentrated and aligned to aid polymerizations. He pointed out that specific isomers (18) of amino acids and sugars have affinity for some crystalline surfaces, and speculated that this may be the reason why all living cells use only the left-handed isomer of amino acids, as well as only the right-handed isomers of sugars, although both isomers are found in abundance in the environment. However, crystalline structure is by its very nature simple and <u>repetitive</u> and any polymers so formed would necessarily reflect this.

All living organisms are made up of cells

Robert Hooke, in the 1660s, magnified thin slices of plant material and recognized that they were made of cells. After better microscopes were invented, Theodor Schwann in the 1830s called the cell "the unit of life".

There are two kinds of cells today, prokaryotic and eukaryotic. This division is based on whether or not they have internal phospholipid membranes. The prokaryotes have only the outer boundary membrane, while eukaryotes, in addition to the outer membrane, have several internal, membrane-bound organelles. (Pro means before, and karyote means kernel). The prokaryotes are further divided into two categories: the archeobacteria (Archaea), and the eubacteria.

In recent decades, biologists have revised the system of classifying all living organisms (19). The highest division is now called "domain". There are three domains recognized: archeobacteria, eubacteria and eukaryotes. All plants and animals are classified as eukaryotes. The first two domains consist of single celled prokaryotes and they are structurally the simplest cells on earth today. *Escherichia coli,* the organism studied first by so called, molecular biologists, is a eubacterium. It is wonderful that the genetic code originally formulated using *E. coli* is the same in all organisms now studied.

Archaea

As implied by their name, the Archaea are believed to represent the most ancient forms of life on earth. They are morphologically similar to eu(or true)bacteria, but the two groups have some important differences: Archaea and Eubacteria use differently sized ribosomal RNAs, and they differ in the arrangement of their DNA in that only Archean DNA is involved with histone proteins (20). In these two fundamental aspects, the Archaea resemble Eukaryotes (i. e. all other organisms, including ourselves) more than they resemble eubacteria.

Suggesting their primordial status, most Archeea, unlike eubacteria, obtain energy from inorganic sources, such as hydrogen sulfide and metal ions, that were common in the environment of ancient earth. (Incidentally, oxidation, was impossible then because there was practically no free oxygen, most of it being locked up as parts of minerals.) Also most Archaea are more tolerant than other living cells of extreme environments of heat, salinity and pH, which environments were probably common on ancient earth. (21).

Some Archeans have also been found in our own intestines.

Eubacteria

Eubacteria, beside having massive influences in our modern environment (not just in disease, but in essential recycling of elements), are now also believed to have formed crucial parts of eukaryotes: Their descendants are the vital eukaryotic cell organelles, mitochondria and chloroplasts.

Decades ago Dr. Lynn Margulis (who was a professor at UC Berkeley and at the University of Chicago, as well as spouse of astrophysicist, Dr. Carl Sagan) proposed that mitochondria and chloroplasts were derived from ancient eubacteria, that were edocytosed (a cellular ingestion process) into archeobacterial-like cells. The Margulis theory has been substantiated by the discovery, within modern mitochondria and chloroplasts, of remnants of bacterial protein synthesis machinery, including DNA molecules (22). Thus the eukaryotic cell appears to be the result of cooperation between Archean and prokaryote ancestors.

Cellular architecture

Several physical forces are important in the production of the amazing "architecture" of living cells. Forces which are not directly governed by the genetic code are reviewed by Franklin Harold (23). To quote his article: "directional physiology, spatial markers, gradients, fields and physical forces" [including gravity] play a role." Harold concisely states: "spatial organization is not written out in the genetic blueprint [but] it emerges epigenetically from the interplay of genetically specified molecules." Cell organization appears also to be partly dependent on structural modeling, as shown by observations that cell walls in bacteria require that at least parts of these structures be present, before new construction can begin .This to me recalls the statement made by Rudolf Virchow in the latter part of the nineteenth century: "All cells come from preexisting cells".

RNA world

Messenger, transfer and ribosomal RNAs are essential in cellular protein synthesis machinery (11). In addition, newly discovered

RNA molecules that have different (mainly regulatory) functions have been described (24).

Sczepanski et al have shown that small synthetic RNAs are capable of replicating themselves, in vitro i.e. without any enzyme catalysts (25).

Double stranded RNA (dsRNA) exists in some viruses (26), and dsRNAs can be replicated directly (27), but these viruses still must rely on their host's protein synthesis machinery to make proteins.

All this has led scientists to propose that life may have originated with an RNA molecule, but the synthesis of proteins, which are essential elements of life, has not been addressed to date in this "RNA World" hypothesis.

I have tried to picture a prebiotic RNA world: Relatively, simple RNA molecules, with repetitive subunits could have been contained in concentrated islands, perhaps membrane bound. Such islands could have replicated by fragmentation, but they probably could not have survived unless they also had some source of energy. Such energy could have been supplied by a simpler version of the Modern Krebs cycle, which in most scenarios uses energy from inorganic sources in hot acidic water, (see references in 13). This self-replicating cycle, relies on several, specific organic molecules that would have had to have been subsumed by those prebiotic islands.

But without protein enzymes to efficiently catalyze reactions at life-permitting temperatures, further evolution of structure and function is hard to imagine.

The Genetic Code

I have tried to show that life is dependent on the structures of proteins, which in turn rely on the sequences of different amino acids of which proteins are made. These sequences are formed using a genetic code, which is essentially the same in all life as we know it.

Gregor Mendel made a monumental breakthrough in our understanding of the laws of inheritance in the nineteenth century (28). Then in the early twentieth century observations of colored bodies (literally chromosomes) became possible because of better microscopes, differential staining (29) and eventually using living cells in culture under phase contrast microscopy, coupled with time lapse photography. The behavior of chromosomes fit Mendel's laws that defined the behavior of his factors (later called genes).

Earlier, toward the end of the nineteenth century, Friedrich Miescher showed that the material found in nuclei of white blood cells contained nucleic acids. Up until the middle of the twentieth century, it had been thought that if any chemical was responsible for heredity, it must be a protein, while nucleic acid molecules (mainly DNA) were probably only structural. (I even recall an early textbook, which said so.) After all DNA is rather uniform, composed of only four, similar organic bases, one kind of sugar and some phosphates. In contrast proteins, which are found everywhere in cells, have at least twenty differently shaped subunits (the amino acids), and they perform many functions, from digestion to muscle development.

The truth was understood only after James Watson and Francis Crick elucidated the structure of DNA in the 1950s (30). Watson and Crick's seminal paper on the structure of DNA concludes with this modest sentence: "It has not escaped our notice that the specific

11

pairing we have postulated immediately suggests a possible copying mechanism for the genetic material." That publication was followed by a flurry of reasoning and experiments, which showed that the genetic code is written in triplet words (an idea first proposed, by George Gamow, a Ukrainian scientist who among other things coined the term "Big Bang") on the basis of simple math: there are four letters in DNA and twenty "words" are needed: Two bases at a time could only make twelve words, but combinations of three bases would suffice.

Elucidation and isolation of the machinery used by cells to make proteins allowed investigators to perform experiments in test tubes: RNA strands, natural as well as artificially synthesized, were used to make proteins or homopolymers of amino acids in vitro. Many scientists were involved, including Marshall Nirenberg, Philip Leder and Seymour Benzer. Synthetic RNA polymers were used to find which RNA triplet codes for which amino acid: e.g. Niremberg started with the homopolymer of uridine, and found it produced only poly-phenylalanine.

Further direct proof of the triplet nature of the code came from experiments by a team that included Francis Crick (31). These ingenious experiments involved deleting bases: When one or two consecutive bases in a natural RNA strand were deleted this always disrupted the production of that protein. However, deleting three consecutive bases restored that protein's production. It was reasoned that the third point mutation put back the reading frame of the protein synthesizing machinery.

This linear genetic code (figure below) is shared by all of life on earth, from the most ancient bacteria to the largest vertebrates, organisms found in all environments from the depths of the oceans to the highest mountains (32). In this chart, nitrogenous bases are designated by capital letters (A, U etc) and amino acids, for which they code, are designated by the three starting letters of each amino acid name (Phe, for phenyl alanine, Leu for leucine etc.)

Second letter

First letter		U	C	A	G	Third letter
U		UUU] Phe UUC] UUA] Leu UUG]	UCU] UCC] Ser UCA] UCG]	UAU] Tyr UAC] UAA Stop UAG Stop	UGU] Cys UGC] UGA Stop UGG Trp	U C A G
C		CUU] CUC] Leu CUA] CUG]	CCU] CCC] Pro CCA] CCG]	CAU] His CAC] CAA] Gln CAG]	CGU] CGC] Arg CGA] CGG]	U C A G
A		AUU] AUC] Ile AUA] AUG Met	ACU] ACC] Thr ACA] ACG]	AAU] Asn AAC] AAA] Lys AAG]	AGU] Ser AGC] AGA] Arg AGG]	U C A G
G		GUU] GUC] Val GUA] GUG]	GCU] GCC] Ala GCA] GCG]	GAU] Asp GAC] GAA] Glu GAG]	GGU] GGC] Gly GGA] GGG]	U C A G

Why the genetic code is an argument for creation

As discussed above, formation of organic molecules and their polymerization, as well as formation of membranes and the present architectural complexity of cells, all could have been guided by physical forces alone. Also, the present complexity of today's living organisms is evidently the result of Darwinian natural selection. However, I claim that the genetic code, which is basic to all life as we know it, could not have happened by the process of evolution because for this process to function each change must have selective advantage. But there is no advantage to a particular codon word being associated with its specific amino acid, and having only one or two of the codon words does not produce useful sequences.

The genetic code is universal and unique, i.e. all living cells on earth use the same code, from the most primitive bacteria to large mammals, like us. Natural selection could not have produced this complex, arbitrary code that has spread, with almost no modification (32), to all niches of life on earth during 3.5 billion years. In other words, the code must have been "nearly perfect" to begin with, and started only once.

Did God speak three letter words in the afternoon of the third [so-called] day, described in the Bible (33)?

References and Footnotes

(1) This is the Wikipedia definition of chemistry: "The branch of science that deals with the identification of the substances of which matter is composed."

(2) The idea of covalent bonds, i.e. the sharing of electrons between atoms, was introduced by Irving Langmuir in 1919. Although the element most often involved in covalent bonding is carbon, other atoms (mainly hydrogen, nitrogen and oxygen) can also share electrons. Covalent bonds involve electrons in the outermost shell of atoms.

(3) See Miller/Urey Experiment in *Wikipedia*.

(4) In large organisms structural proteins form anatomy such as muscles, hair and connective tissue, and in individual cells they help hold together some structures, e.g. ribosomes.

(5) Frenkel-Pinter, M et al *Selective incorporation of proteinaceous over nonproteinaceous cationic amino acids in model prebiotic oligomerization* reactions; <u>PNAS</u> 116 (33): 16338-16346; (2019).

(6) To illustrate the flexibility of proteins, consider diseases caused by prions. Prions are miss-folded proteins with the ability to transmit their miss-folded shape onto a related normal protein. They characterize several fatal and transmissible neurodegenerative diseases in humans and many other animals.

(7) Articles explaining *Hydrogen Bonding and Van der Walls Forces* are cited in <u>Biology Libra Texts</u> (2019).

(8) The nitrogenous base, thymine and the sugar, deoxyribose are found in DNA. While the nitrogenous base, Uracil and the sugar, ribose are found in RNA.

(9) Cafferty, BJ et al. *Spontaneous formation and base pairing of plausible prebiotic nucleotides in water;* <u>Nature Communications.</u> 7, Article 11326 (2016).

(10) Hazen, R *Origins of Life,* The Great Courses, Chantilly, Virginia (2005).

(11) The modern cell's protein synthesis process is called translation. It depends on many enzymes as well as on three RNAs: messenger, or mRNAs (which transmits the genetic code message), transfer, or tRNAs (which translate each triplet codon into its corresponding amino acid) and ribosomal, or rRNAs (which form ribosomes, a kind of "work bench" for this process). Translation can be made to work in vitro, with isolated components of the cellular machinery.

(12) DNA's backbone, in contrast to RNA, is not flexible because of one small repeated side-group difference in the backbone of DNA. (As stated in the text, messenger RNA does not fold, which makes it easily degraded, unless it is protected by specific transport proteins.) Transfer RNAs form shapes that display the anticodon triplet for a particular amino acid at one fold, while the specified amino acid is attached at an opposite end of the tRNA molecule. (This attachment is governed by appropriate enzymes.) Ribosomal RNAs (there are two, each having a different size or weight) combine with several structural proteins to form the ribosome, which provides the surface on which protein synthesis takes place, under the direction of several enzymes.

(13) Stubb, RT et al. *A plausable metal free ancestral analogue of the Krebs Cycle composed entirely of alpha keto acids.* <u>Nature (Chemistry)</u> October (2020).

(14) Energy is stored in high-energy, covalent bonds between phosphates of ATP (adenosine triphosphate) and GTP (guanosine triphosphate). The greatest amount of energy (approximately 7 kcal/mole) is in the bond between the second and third phosphate groups. This covalent bond is known as a pyrophosphate bond.

(15) See Micelles in *Wikipedia*. Micelles form only in a polar medium, like water. Does that show that life must have evolved in water?

(16) Hazen, R Origins of Life, Lecture 15: *Lipids and Membrane Self-Orgnization;* Published by The Great Courses, Chantilly, Virginia, (2005).

(17) Ibid. Chapt.15 *Macromolecules and the Tree of Life.*

(18) Isomers: Because of their inflexible shapes, organic molecules display handedness (analogous to our left and right hands) or Isomeric forms. Pairs of isomers behave identically in simple chemical reactions, but both cannot be utilized by living cells, e.g by enzymes, and do not get assimilated. (BTW: The diet industry takes advantage of this and flavors foods with left handed sugars.)

(19) The living world is classified into a nested series of categories. Domain is the highest, or largest, category. So to illustrate: we, humans, are in the same Domain with every other eukaryote, including e.g. Amoebas (but excluding the Archaea and prokaryotes); then moving down we share the same sub-Phylum with all animals with backbones, including fish, and we share the same Class with all other mammals e.g. cats, but we share the same Genus only with other, now extinct, humans. (See The Three *Domain System* in *Wikipedia.* The term domain" now replaces the old term, "kingdom")

(20) Histones are proteins that are important in folding the long DNA molecules into compact units called nucleosomes. Interestingly, histones are found associated with DNA in Archaea but not in eubacteria which have a single, circular and naked DNA molecule.

(21) Part of their adaptation to extreme temperatures Archeans have a DNA polymerase, which is very useful to today's molecular biologists: This enzyme remains stable even in high heat and therefore functions well in polymerase chain reactions, which require cycling in an out of hot temperatures, and which are crucial today for sequencing DNA.

(22) Modern mitochondria and chloroplasts may have retained some uses for the DNA they contain. However, most of the DNA instructions for the construction of mitochondria and chloroplasts are found in the cell nucleus. It is interesting that the structure of the DNA found in mitochondria resembles prokaryotic DNA in that it is circular and is not packed by histone proteins. Other relevant features of mitochondria have been reported recently: Boguszewska K et al, *The Similarities between Human Mitochondria and Bacteria in the Context of Structure Genome, and Base Excision Repair System Molecules.* 2020 Jun; 25(12): 2857. Published online, June 21, (2020) 25122857 PMCID: PMC7356350.

(23) Franklin MH, *Molecules Into Cells: Specifying Spatial Architecture;* Microbiology and Molecular Biology Reviews, Dec (2005).

(24) Clancy, S; *RNA Functions* Nature Education (2008)

(25) Sczepanski, JT and Joyce GF *A Cross-chiral RNA polymerase ribozyme* Nature October 39 (2014).

(26) *Double stranded RNA Viruses,* Wikipedia

(27) The sequence of bases in nucleic acid polymers (whether DNA or RNA) is always copied directly (with appropriate synthetase enzymes) . To replicate, nucleic acids must be double stranded, with the two strands being mirror images of each other.

(28) Gregor Mendel's report (1885) about inheritance in pea plants was published in a botanical journal and at the time largely ignored. It was rediscovered fifteen years later by at least

three scientists: Hugo DeVries, Carl Correns and Erich von Tschermak, each working inpendently in different European countries. Confirmations of Mendel's data spurred many investigators to further experiments. The most notable of which used the fruit fly, *Drosophila melanogaster* in the laboratory of TH Morgan at Columbia University in New York, in the first half of the 20th century.

(29) *The Man Who Invented the Chromosome: A life of Cyril Darlington,* Heredity, 97:136 (2006).

(30) Watson ,J and Crick, F *Molecular Structure of Nucleic Acids; A Structure for Deoxyribose Nucleic Acid,* Nature, April 1, (1953)

(31) Yanovsky, C, *Establishing the Triplet Nature of the Genetic Code* Cell: 128 (5) 815-818 (2007)

(32) Only a few and minor differences in the genetic code have been found: some small parasitic microbes, called mycoplasma, incorporate a couple of amino acids other than the basic twenty into their proteins. Also slight differences in the code for certain amino acids have been found in some microbes as well as in the "vestigial" DNA found in mitochondria.

(33) Genesis, Chapter 1:11 "Then God said: *Let the earth bring forth vegetation.."*

2. The Wonders of Water

When cosmologists find liquid water on other planets or in their moons, it suggests that some kind of life might exist there. For example, Jupiter's moon, Europa, is of interest, even to science fiction writers (1), because it was found to have a liquid water sea locked under its icy surface.

Water is essential in many processes of life. It is a remarkably stable molecule, made up of two hydrogens and one oxygen, bound together covalently (i.e. the atoms share electrons). Whenever bonds between atoms are broken energy is released. (2).

In photosynthesis green plants focus light to break up the bonds in water. This process (3) starts when chlorophyl (with the help of accessory pigments) absorbs light energy. This energy is passed on within the chloroplast, with the help of protein enzymes, through conjugated double bonds, until it specifically breaks the bonds between the hydrogen and oxygen of water. Two important products of this process are molecules of ATP (4) and the production of NADPH, whose hydrogen is said to be "exited", i.e. ready to combine with other molecules (5).The reactants involved in these, so called light reactions of photosynthesis, are held in precise relationships to each other by membranes. The "excited" hydrogen, is then passed through another cascade of coenzymes to be applied by the dark reactions of photosynthesis to the synthesis of glucose sugar (6).

NADPH is the final coenzyme product of the light reactions. It holds energy as an excited hydrogen, which is then guided by another series of enzymes,to deliver its excited hydrogen atom to a molecule of CO_2.With this energy the carbon atoms in CO_2 link together to make the sugar, glucose. These anabolic reactions, or the synthesis of sugar are the second phase of photosynthesis, called its "dark reactions" (6).

Sugars can release the energy stored in their covalent bonds and they can be converted into other molecules required for life. All these reactions require specific protein enzymes, and the entire system of energy release and molecule synthesis is called metabolism (7).

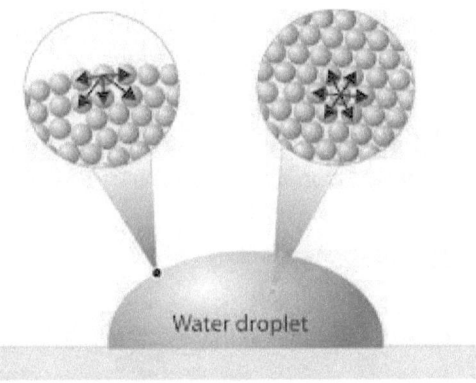

Water droplet

The role of water in photosynthesis is: just one of water's wonders: Life is supported by water's other unique physical attributes. Here are four that come to my mind:

- The weak, so called hydrogen bonds between water molecules, myriads of which make up even the tiniest drop of water, produce surface tension and cohesion (see diagram). This makes it possible for water to travel from the roots to the leaves of even the tallest trees. (8), and wispy animals, like water striders, can walk on water.

- Both the liquid and solid states of water exist within relatively mild temperatures. Water weighs less when it is solid than when it is liquid, because the crystalline structure of ice packs the molecules of water further apart than they are in liquid water. This allows creatures dwelling in fresh water to escape being imbedded in a solid, when the weather freezes.

- Many molecules dissolve in, i.e become separated by, water which enables interactions with neighboring molecules.

- Guided by specific enzymes, water can modify organic molecules by being inserted into their structures or by splitting their organic bonds.

I'm sure there are other important functions that water performs to sustain life. After all our bodies are more than half made of water.

References and Footnotes

(1) For example, the film, *2010: The Year We Make Contact* - Wikipedia

(2) We get energy from what we eat because the bonds in the organic molecules in our food are severed by many enzyme-driven reactions. The final step of this so called metabolic process is accomplished by the cytochrome system, which is housed in the membranes of mitochondria.

(3) *The Light Reactions of Photosynthesis* were elucidated by Daniel Arnon at the University of California in Berkeley. PNAS **66** (11):2883 (1971).

The co-enzymes of the light reactions are held in place by membranes within chloroplasts, which facilitates the cascading transfer of energy. These systems of membranes within chloroplasts are called grana.

Chloroplasts, as well as mitochondria. are probably descendants of ancient prokaryotes. See a discussion of prokaryotes in the previous essay on the origin of life.

(4) ATP is discussed under the heading of <u>energy</u> in the previous assay of the origin of life.

(5) Nicotine dinucleotide phosphate is a coenzyme that accepts an energized hydroden to become NADPH.

(6) Nobel Prize winner, Melvin Calvin, who was also a professor at UC Berkeley, worked out the dark reactions of photosynthesis by using radioactively tagged starting reagents. These synthetic reactions are called dark because they occur after the capture of light is complete.

(7) When I was a student we decorated our walls with a huge chart, entitled "Intermediate metabolism". No specific organism was cited in these charts, because <u>most</u> of the basic reactions shown in these charts are shared by all living cells. A glaring exception to that last sentence is, of course, that only green plants can photosynthesize. Moreover various organisms may lack one or another of the needed metabolic enzymes. These "deficiencies" probably reflect the environment in which different organisms evolved: Why make enzymes to drive syntheses of products that are readily found in one's diet?

(8) The truly astonishing, daily feat of water moving from roots to the crowns of tall trees is thought to occur because of the cohesion of water molecules coupled with the evaporation of water from leaves, which pulls a column of water in the narrow channels of a plant tissue, called, xylem.

3. To Shed Some Light

On the difference between human and other animals

Morality, i.e. the ability to distinguish right from wrong, is thought to be uniquely possessed by people. So when non-human animals do something that disturbs us, like my well-fed cat killing a song bird, we say he is not to blame because he is only acting out of instinct. Instincts should not be confused with reflexes that involve only quick, momentary nerve impulses, and bypass the brain. Instincts are much more complex.

Since carnivores in the wild must kill to get nourishment, their hunting abilities have selective value. This is obviously true in regard to physical features like teeth and claws, but what about their hunting skills that need to be taught to carnivores, such as young lions and fledging eagles, by their parents (1)? Evolution can only work with traits that are reliably heritable. Can linear DNA code for hunting strategies, which can be quite complex, such as that found in wolf packs (2)?

Do people have instincts? Is the human need for revenge an instinct? When someone hurts us, a need to retaliate often persists even when we realize its futility, and when we know that such fuming can lead to harmfully high blood pressure. Successful coping depends on us learning to channel such feelings into constructive action. This skill depends on reasoning and planing, and this in turn may only be possible with some sort of language.

All animals communicate with behavior and with sounds. Even my cat's meow clearly varies. Bird songs have been found to signify things like,"this is my territory" or " I am looking for a mate". Great apes in captivity have been taught to use sign language, but their messages appear to never rise above expressions of simple needs or desires. Although some, apparently sophisticated non-human communication has been revealed by astonishing studies with corvids that demonstrate ability for memory and planning: Crows have been shown to pass on recognition of a new danger to their offspring, even long after the particular danger is passed (3).

However, human communication is still beyond that of any known in non-human animals: By reasoning, people can envision new future scenarios, and combine different observations and abstractions. All this involves symbols and words, arranged in syntax. Perhaps the purest illustration of this is the use of higher mathematics in physics. It is astonishing that quantum physics, whose application makes the modern digitalized world so successful, can at present only be clearly described by equations.

So far all the communications we have considered are utilitarian, but what about aesthetics? What is beauty and why do we appreciate it? In this context students of animal behavior have pointed to bowerbirds, who elaborately decorate the entrances to their nests. Although this functions to attract a mate, and therefore it is still utilitarian, it rises above things like beautiful plumage, because the bowerbird selects and arranges various, mostly blue objects.

Most people value seeing or hearing something beautiful, whether it was made by nature, like a flower and a sunset, or by humans like art and music. It is like love we cannot define, but need. Evidence for such a need has been found in all human societies. Cave images made with red ochre, shell jewelry and engraved stones have been dated as far back as 100,000 BC. While admirable animal dwellings, such as beehives, always have an essential use and are never just for decoration.

Human ability and creativity in art, as well as our extensive use of tools, are aided not only by our brain development, but also on our opposable thumbs and upright posture.

Asking whether these anatomic "innovations" became useful because of our complex cognition, or whether we developed our brain power because of them, seems futile because evolution occurs stepwise in stages, and therefore which factors are the most selective may vary with different circumstances.

Is love, when not easily linked to reproduction, unique among humans? Our pets seem to demonstrate loyalty, and some wild animals appear to show such attachments: It is hard to see the grieving of elephants over the death of a companion (4) as adaptively useful.

On the other hand, supposed altruism demonstrated by prairie dogs, who alert their packs when danger approaches (5) is clearly adaptive because it benefits a closely related prairie dog population. Is human altruism, only governed by our selfish genes (6), or does saving an unrelated someone from pain, rise above this ?

Most people think that only humans are sentient, which may be defined as "understanding ideas in compact condensed forms".

Thus killing non-human animals, as long as it is done without causing them pain or fear, is excused because they are not sentient. Some non-human behaviors suggest that the difference between non-human animals and people is only quantitative. This notion conforms to an idea basic in evolution, i.e. that change occurs in degrees. Teilhard de Chardin, who was a Catholic priest and also a geologist, formulated this philosophy over a hundred years ago (7), and saw no conflict with it and his belief in God, who Teilhard saw is continually drawing creation to His level.

References and Footnotes

1. Morell, V *Do Animals Teach* Natl. Wildlife; Oct-Nov, (2014).

2. Escobedo, R et al *Group size, individual role differentiation and effectiveness of cooperation in a homogeneous group of hunters;* JR Society Interface, (2014).

3. *Crows share intelligence about enemies;* CBC News June 30, (2011).

4. Satin,C *;Depths of Animal Grief;* Nova, July 8, 2015.

5. Hoogland, JL *The black-tailed prairie dog: Social life of a burrowing mammal,* The University of Chicago Press. (1995).

6. Dawkins, R *The Selfish Gene,* Oxford University Press, (1976).

7. Teilhard de Chardin, P *The Phenomenon of Man,* translated by Wall, Bernard. New York: Harper (2008).

4. The Theological Virtues

"The theological virtues….are infused by God into the souls of the faithful.."(1)

.."faith, hope and love remain, these three; but the greatest of these is love." (2)

<u>Faith</u>

Moments after my husband died, I phoned my sister, who was living far away. "Did you open a window?", she asked.

"I'm opening it now."

Five years later, when my sister lay dead in a hospice room, I made sure a window was open. It was comforting to think that I was obeying her wish.

Why was that window-opening custom reassuring? After all, quantum physics claims that fundamental particles, can penetrate walls, and what could be more fundamental than the stuff of souls? Why did I think that it was good to make the pathway for souls as easy as possible? Why do I believe that human beings have immortal souls, anyway?

My answer to these questions is: "because of the theological virtue of faith."

We operate everyday using at least a modicum of faith. This kind of faith is better called trust. We readily fall back on trust in making small decisions because we have other more important things to do than to examine every proof. For example, we trust that the products we buy are as advertised, or that civic ordinances and laws are instituted for our good. We probably believe that this trust can and will eventually be tested by consequences, i.e. if we are fooled, the perpetrators will be exposed.

We also tend to have faith/trust in various professionals since none of us can be experts in every field. Still even here, we can (if we take the time and trouble) research and verify their assertions. But can our faith in God be justified only after death?

Here are some sources that to me provide some justification for theism:

We have the testimony, consistency and perseverance of saints, some of whom were willing to suffer and die rather than denounce their faith. Sometimes, we may sense God's presence in uncanny events that seem unlikely to be just coincidences (3). We also have the writings (and today videos) of theologians, who give us well-reasoned dialogs, and we have God's direct words in scripture, which when studied ring true because of their simplicity, profundity and consistency. Finally, most important to me, we know faithful people personally, whose actions and words demonstrate their faith. Although God is above and beyond the world as we know it, true believers display a "peace that is beyond understanding" (4).

Hope

If we lose hope we fall into despair and cease to really live. This was the cardinal sin of Judas Iscariot. If he had only reached out to Christ (even at the foot of the cross), he would have been forgiven and given us, myriads of subsequent sinners, an extra dose of hope. When we stop hoping we stop working to change things and consequently all the worst things we can imagine will happen. All I can say is, *thank God for Hope.*

Love

Even the most fervent agnostics acknowledge the unique importance of love. No one has described it better than St. Paul (5).

Some maintain that love is only the result of evolution because it aids survival and reproduction, or that shielding selected individuals was selected by evolution because it aids the tribe, whose members are genetically related. But it is difficult to explain altruism showered on distant tribes or even on different species. Moreover, utilitarian thinking can't explain why we love beauty in nature (6), art, architecture, music and words. We love certain people for their inner beauty, for what they say and do, for the ideas they pass on, and for their reaching out to make the world happier and healthier for everyone.

Love is something we need, but can't define. It is often said that God is Love. We can't really explain either completely.

In my experience, human love can make us feel grounded (7). Does God's love make us feel grounded? Theological treatises, perhaps especially those of mystics can help, but we need God's direct help to start. It is a theological virtue.

God gives faith, hope and love to all who desire them with a sincere heart. They are like seeds, and like all seeds in order to grow they need good soil (our sincere heart), water (institutional sacraments) and sunshine (help of loving people).

The three theological virtues are part of God's care for us, but God always expects us to do our part to make them grow (8).

References and footnotes

1. Catechism of the Catholic Church; article 1812; <u>Doubleday,</u> New York (1995).

2. 1 Corinthians, 13:13.

3. The day my husband died and after I called my sister, I went for a walk in a large nearby park. It was the middle of a weekday and the park was deserted. Somehow a man started walking with me. At other times I might have felt nervous at the intrusion, but his presence did not cause me any anxiety. To the contrary, his soothing words calmed me. I remember him saying: "The kingdom is within you," and this made perfect sense to me at the time. As we departed I invited him to a memorial I was planning, but he said he was too busy to attend. I smiled, sure that he was an angel in charge of grief counseling.

4. Philippians 4:7

5. 1 Corinthians, chapter 13.

6. When I was an undergraduate, the priest in charge of the Newman Club told us a story of a very poor family he knew that lived in another country. The mother hung up on the wall of her little hut, a bright photo of a red tomato that she cut out from the label on a soup can. That little bit of beauty made her family smile. So it is with everything beautiful we see, hear or understand.

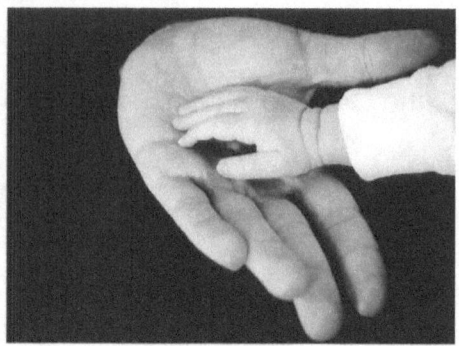

7. We brought our son home, three days after his birth, while I still had a slight ache in my belly. We were then living in a small, walk-up apartment. I gave our precious bundle to my husband saying, "put him in his little crib and then come back to help me up the stairs." Many women feel post partum depression, due to hormone changes, but I felt grounded because of my husband's gentle support and smile that radiated love.

8. Mark 4: 1-20; Matthew 13: 1-23 and Luke 8: 4-15, *The Parable of the Sower*

5. The Birth Control Problem

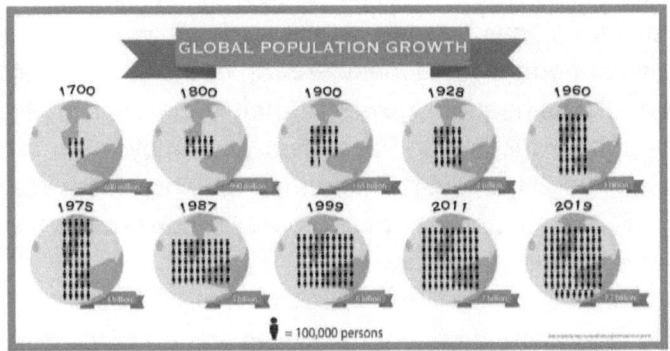

I am trying to be a sincere Catholic, but I balk at being told what is right without any discussion. This is true when it comes to the Church's stand on birth control. Therefore, I hope to provide some discussion about this important issue here, because control of human fecundity is essential to sustain our beautiful world. (Witness recent devastating news about the destruction of the Amazon rain forrest that has resulted from too many people burning the forest in vain hope of gaining arable land.)

In the 4th century, St. Augustine declared that the purpose of marriage was to produce and raise children. It seems to me that the Catholic church has promulgated this, to my mind a narrow idea, ever since, saying something like this: "The use of any devise that would obfuscate God's intent to create a child during the sexual act, is sinful."

Birth control, although practiced from time immemorial, became more practical when modern methods (primarily "the pill") became available. These means are used today in industrialized countries by a most married people, and I don't think there is any conclusive evidence that their use promotes extramarital sex. After all, prostitution has been called "the oldest profession". A commission was set up by pope St. John XXIII in 1963, which included high-level clerics and theologians as well as married, lay people to study this issue, concluded that a reconsideration of the church's stand on birth control was needed. However, their suggestion was ignored by Pope John's successor, Pope Paul VI, and by the church ever since. Still I think that even the staunchest church conservatives would agree that using devices to prevent conception are better than using abortion to control birth.

The only relaxation in the church's teaching on birth control, has been its approval of he rhythm method (1), which relies on monitoring female hormonal cycles to determine a time when ovulation and therefore fertility is present. This method is about 80% effective, and it can be used effectively only when a woman's menstrual cycles are regular. (I think that the church's upper hierarchy has said that the use of oral hormones is permitted to regulate menstrual cycles, *if deemed medically necessary.*) It is also noteworthy that even the 1963 commission (cited above) stated that the use of the rhythm method can strain a marriage because it requires cumbersome measurements. However, I acknowledge, that discussing the wife's needs, such as is necessitated to monitor her cycles, can enhance love between couples.

Some people assert that "artificial birth control" may harm women's health. Admittedly, there is risk of potential harm in all medical interventions and contraceptives are no exception: hormones in "the pill" may increase the risk of cancer; intrauterine devices may damage the uterus; barrier methods (condoms or diaphragms) may destroy the spontaneity and joy of sex, and severing the narrow ducts, through which gametes must pass, are surgical procedures with inherent risks. Constant research is improving the safety of these so called artificial contraceptive methods. In comparison health problems resulting from pregnancy are much more common than those resulting form any artificial contraceptives.

I maintain that it is a moral obligation to limit the number of our offsprings. Some may say: "If you can support a large number why not have many children?" Even disregarding the fact that childbearing is a possible health burden for women, and that women as well as men have the need to engage in activities other than bearing and raising children, the undeniable fact is that no matter how much you support your children's well being, including their education, each added person uses nutrients and oxygen, produces carbon dioxide, sewage and waste and takes up space. The world will become an uncomfortable slum if human population continues to grow unchecked, as is presently the case in most of the world. Do we want to wait until human numbers are drastically checked by disease?

Even really good people can completely ignore this reality: A priest friend of mine once told me how his sister, a missionary nun in Africa, wrote to him about elephants trampling their vegetable garden.

"At least that's one thing we don't have to worry about here," he said smiling.

"Well", I said, "that vegetable garden probably diminished those elephants' territory." The point is that the world has limited space, and the natural world depends on a balance of many interacting species. So too much of any one species is a road to disaster. A most graphic, historic example of this is the Irish potato famine (2).

Preserving our natural world, with its marvelous fauna and flora, is to me second in importance only to believing in God. I think evidence plainly shows that human beings have been changing the natural environment for the worse in many ways. We are in the midst of an ongoing extinction of plant and animal species at very hight documented rates. Our survival and the survival of the natural world depends on having a variety of interacting species (3). This present extinction is not due to an asteroid, to massive volcano eruptions, or to any cause over which we have no control, but it is happening because of human activity, including manufacture of polluting gasses, changing green areas into deserts and people arrogantly wasting natural resources.

Is the use of artificial contraceptives the only solution? It is clear that the size of families is diminishing in most countries with a "high standard of living". The very poor need their children's help and labor, and for such poor, children may be the only positive gift they can see. If the standard of living for all people were raised throughout the world, would the world population (4) stop its increase?

References and Footnotes

1. Gorringer, D *How far did they go? Challenging assumptions about Catholic women and the pill* The Tablet. (2019)

2. Bartoletti, S *Black Potatoes: the Story of the Great Irish Famine;* Houghton, Mifflin and Harcourt Publishing (2001).

The potato originated in South America where it flourishes in a number of varieties. Sir Walter Raleigh brought one of these varieties to Ireland in 1589. In 1845 a plant disease wiped out all potatoes there. While it is true that Ireland's humid climate was a factor, scientists agree that had there been more varieties some would have survived.

3. Barnosky, AD; *Dodging Extinction: Power, Food, Money and the future of Life on Earth,* Paperback; University of California Press (2016)

4. Population Growth, Wikipedia

6. How Evil and Power Changed My Family

My father in Prague (left photo) and a refugee (right photo)

<u>Preface</u>

Prior to World War II, Czechoslovakia, my native country, had been growing into an ideal democracy, but after the war it became a repressive Soviet dictatorship. My father, Judr. Svatopluk Boruvka, wrote the diary printed below, in the spring of 1949. I found it among our family papers. It chronicles his escape from the land of our birth, and the hardships he endured immediately thereafter.

Father addressed this diary to my mother, his wife, Pavla. I suspect the original was in Czech and was translated into English by my sister, Shari Boruvka-Roth, whose given Czech name was Sarka, and who in late life became a writer.

I made only slight changes in father's narrative, to smooth some of the text. I also retained some of the names of people mentioned therein, because I reasoned that this can have no consequences now. Finally, I added three small sections, which are indicated by being indented: The first describes the sham election of 1947, the second tells how father was taken into custody for questioning and jailed for days in Prague, without ever being charged, and the third outlines events that happened after Father's diary ends.

Frankfurt, May, 1949

I fervently hope and pray that you are reading this in peace and safety. I am writing this for you, dear Pavlo (1), and for our two little daughters so that you will know how it came to be that we had to separate so suddenly, and how it was with me after that.

Political Background

It hung in the air for a long time. We should have recognized its coming, for many events should have been clues. We realized too late that it was not just a serendipitous improvisation, but a careful plan, drafted in Moscow during the war. The Soviet Communists must have counted on our thinking of them as Russians, our Slavic brothers. This was true of course. (Remember how you took particular pride in learning to speak Russian?) We dismissed condemnation of Soviet Communism, thinking it to be a distant experiment, never believing that Stalin could influence a revolution in our country. I thought that the leaders of our newly born democracy would know better than to cut from under themselves the branches on which they were sitting.

I have to give the insurgents credit for one thing. They displayed great discipline. No Communist that tended to "emotionalism", which is how they labeled having compassion, was allowed to function. Anyone who openly showed such "weakness" was eliminated. For example, Duris and Kopecky (2) suddenly disappeared never to be heard from again. Eventually, even the famous and powerful, became not immune from annihilation.

In the second half of 1947, the possibility of oncoming disaster became clearer to most of us. Feeble noises made by non-communist parties were allowed only for show. In reality no one dared to say anything against the USSR, and that should have been the most prominent subject of discussion. I attempted to survey news from foreign magazines and newspapers, for example we still had the old "Gunther's Around Europe", and occasionally articles from the "Reader's Digest" were translated and read at our improvised meetings. I am puzzled why at that time my friends and I did not think of joining with some socialist groups. I think we did not understand that movement well.

At first the Communists, clearly instructed by Moscow, sought to advance by specifically ruining the lives of anyone that became "troublesome". If this could be accomplished by simply confiscating property so much the better. For example, without any legal process, they auctioned off Shwarzenberg's estate (3). How brazen the announcements in the paper and on fliers were! It sounded makeshift and homespun, almost like "Kittens for Sale!" There were a few outspoken people like Duris (2), but his small revolt was immediately squelched, and he was never seen again. I, like most people, didn't say or do much, being afraid for my own skin and for the safety of our family. Eventually, even the famous and powerful, were not spared. Didn't they push Jan Masaryk out the window and shoot President Benes in the back of his head? (4)

Svejk (5) would have said "And we, stupid asses, hardly even took notice!" We kept thinking that some sort of justice would be done eventually and that real democracy would prevail. Today, when I read Churchill's Memoirs, which if they could have been obtained then, would have never been allowed to be translated, I can see how naive we were. The mistake possibly lay in that our politicians had little experience and relied on old volumes of political history. They probably held the conviction that their election and education were sufficient preparations for political work. But it is always necessary to keep learning, to observe and to recognize potential insurgencies.

We should have paid more attention to people, who though otherwise politically indifferent, had a healthy sixth sense, and openly dismissed as humbug the slogans being preached by Moscow. Eventually nearly everyone began to see the Communist agitators as a gang of hudlums, when everything kept moving in a wrong direction (6).

The Communist were obviously unpopular and so it was believed that in the upcoming elections they would get, at most 25%. Unfortunately this was just as apparent to the thugs in Moscow and they started more effective actions: They brazenly reorganized the Czech police force, placing only their sympathizers in charge. (Various governmental ministers tried to oppose these changes in vain.) Finally to preclude any possibility that the Communist loose the election they resorted to intimidation. You must remember the phony yes and no cards. They had the nerve to call it a secret ballot !

> For this election the people were provided with only one choice: yes or no to the Communists. All were required to place a yes or a no card in the ballot box, and then hand the unused card to an official, who would consequently know which way the person had voted. In view of the widely known "disappearance" of outspoken people at that time, most feared being labeled a dissident. Similar looking cards featuring a photo of Jan Masaryk, were made available in some circles to enable voters to make a silent protest by using them instead of the yes ballot.

The Communists announced that the election was a landslide in their favor. I wonder if they even bothered to look at the ballots cast.

Most of us expected that the West would be indifferent. We had little hope of their help (even though an internationally staged blockage in Berlin had caused the Soviets to dampen their plans there) because we knew that Western politicians were naturally always led by their own interests. Especially they feared that involvement in Czechoslovakia would mean stepping over an international border and might even ignite another war.

There were no real preparations to stage an opposition against the Communist takeover, and eventually such moves, barely started at the last minute, did not succeed. It's easy to think how stupid we were in retrospect. We pinned our hopes on Jan Masaryk and Benes (4), thinking that all could be saved, convinced that official announcements of president Benes would turn the tide by his authority, not believing that assassination was within the Communist repertoire (7). As if anybody could personally hold back a coup backed by a strong foreign power.

We had not learned lessons from history, a history our nation had lived through before.

Personal danger

At that point I saw clearly that things could become dangerous for me personally. I was a small player in some of the internal politics because of my frequent role as a public defender. I knew that I could never be a Communist or some other opportunist. At that time serious accusations were often made, based on hearsay.

Then any doubts about the precariousness of my position were erased by my lengthy detention for questioning shortly after the Communist takeover.

> I learned, much later, about how father was detained in prison for questioning. During his absence I had assumed he was on some business trip. The Communists kept people locked up indefinitely. I also learned that mother had tried everything to get him released. She even managed to see the wife of president Benes. But this was of no avail.

> Being then eight years old, I was oblivious of what happened the night they came for father, but my sister described to me how two policemen came to get him after dark. Father was out when they showed up, and so they waited for an hour, during which mother pleaded with the stone-faced officers until she was horse.

At that time the authorities used any excuse to eliminate people they deemed to be potential troublemakers. My interrogations in Pankrac (8) revealed to me the instrument they were using:

They latched on to an unjustified claim leveled against me by my niece, who was tied at the time to a man active in a Communist group. I have often tried to understand why she thought so badly of me to actually accuse me of being instrumental in her mother's death in a concentration camp. Maybe it was the fact that I, as a lawyer, had managed the divorce of my brother, who was her father. In reality there was nothing I could've done to prevent the horrible fate of her mother, which she shared with so many Jews.

At the same time, the so-called Practical Independents Party offered me a steady job with them outside of the country. Several in that group urged me to get out while there was still time. I had no reluctance about leaving the country and immersing myself in their work, but I was concerned with the needs of my family. I made unrealistic plans thinking that I could correct my unfortunate circumstances. Some sort of stupid pride would not allow me to abandon my law office, thinking I could live off my savings until I obtained future clients. I considered only my own convictions, paying little attention to those of others. All this mental turmoil, however, suddenly vanished when I perceived that I couldn't stay because it could mean jail and death (9). I still feared how you would react to my leaving, dear Pavlo. Thank God you proved me wrong! You displayed so much courage and common sense! But I am getting ahead of myself.

During late February and most of March, I lived in a strange tension. It was different from the tension prevalent in 1939, before Hitler invaded. At that time I felt being just one among thirteen million others, but this time I knew there were people targeting me directly, people who did not hesitate to use dirty means to get rid of me. Then on March 10, I received the decisive blow: I lost my license to practice law. It is still difficult for me to accept how easily someone's means of earning a living could be taken. All it took was a single official looking letter, no explanations or negotiations were offered. I was permitted to remain in my own office only as a clerk. At that point, my dear Pavlo, we became of one mind that we would be better off abandoning Prague, and we tried to figure out how we could all leave together. To go with my precious family into an unknown world without any plans was unnerving.

I tried to seek help from the lawyers who were now in charge of my office. They suggested I contact Dr. Culik, who had worked with me on several cases in the past. I was told he was not available. He seemed to be on an interminable Easter vacation. I spoke with a kind man who was a clerk in Culik's office, but he could do nothing.

I talked with many people, including a Mr. Vana a who as you know later turned out to be invaluable (10). He was very well versed in various ways, which I could use to leave Prague and assured me they were only slightly dangerous. You and I, Pavlo, spent much time discussing it all, and although we agreed on many things, we couldn't decide what to do.

This shaky situation continued until March 25, on which date in the morning I received a phone call from a certain Mr. Holub stating that he needs to speak to me about something important. We agreed that I would visit him that afternoon. Apparently he had second thoughts, probably thinking it might be bad if I was seen visiting him, for he showed up in our office at half hour before noon. He claimed to have had a conversation with Dr. Konzerzim, who was a presiding judge at that time, which revealed bad things about my case. He emphasized his conviction that it was imperative to be very concerned about the course now in process, that he had heard that some in power said that I must hang! He said that similar proclamations stemming from a Mr. Drda, who had become a Bolshevik Communist overnight and could be very dangerous. Hearing these disturbing allegations, I went to consult a previous colleague, Dr. Johna to learn his opinion and maybe solicit his help. He acted very cordially toward me, but I had the impression the he would not help me. In those days I was not sure who was a true friend or who was ready to betray me. Anyhow I feel it is important to include all these names in my narrative. This probably is a lawyer's habit.

What could have motivated Holub to come to me? I can't imagine that it was out of friendship. After all, he was only a casual, and not a very pleasant acquaintance. I was reminded of that even then because he asked if he could take over our apartment, when we left. He seemed to want to take advantage of whatever he could.

One possible scenario occurred to me: I had been involved in a commercial case during which I had met Holub. The matter involved a sordid, if clever, swindle. Holub would have been the benefactor had the case succeeded and he tried to engage me to essentially facilitate his dirty scheme. He did not count on my figuring it out. I also recall that some of the lines of questioning during my detention in Pankrac mentioned Holub. So this little swindler could have had all kinds of matters on his conscience that made him really scared. Perhaps now he worried that were I to be arrested again, new details could come to light, and so it would be best for him if I were out of the picture. Whatever I thought of Holub, his warning argued strongly in favor of my leaving. Then serendipitously several of my friends, who clerked in the office of the detainee commissioner, delivered an urgent message confirming Holub's assertions: My file was now on their boss' desk. They advised that risking another day was very dangerous and that I should leave the country immediately.

I spent the afternoon and evening of March 25 putting my affairs in order as best I could, and making inquiries about who to contact regarding crossing the border in safety from people I knew in the Practical Independents Party, which now operated undercover and across the border in Germany. I vividly recall my last night with you all. The girls slept sweetly in their little beds, and I fell asleep almost unconscious with exhaustion.

Father's escape

In the morning of March 26 we got up very early. I hugged the girls one last time. Sarka was barely awake, and Dadla was sleeping soundly (11). I did not tell our relatives anything, although I had visited my father the evening before. It was overcast but not cold when I said goodbye to you, by the train destined for the city, Cheb (12). You were so resolute that I felt the weaker one. I had packed only a change of clothes, some food and some jewelry worth about fifteen thousand crowns. Had I known how things would turn out I would have taken much more.

It seemed almost miraculous how easily things went at first. No one took any notice of me. I talked briefly with a fellow who shared my train cabin. He was seeking a business contact in Karlovy Vary. He claimed it was an errand for someone else but I suspected it was for himself. At eleven o'clock we arrived at Falknov, lately renamed "Sokolov" (13).

I had no trouble getting an appointment with my contact, a Dr. Laricky, who invited me to lunch at his home, but seemed somewhat reserved. His wife did most of the talking. I felt he was in a hurry to be rid of me. He proposed that I might be able to accompany a certain physician or veterinarian, who had to travel periodically via a local road to Bavaria. Unfortunately this scheme was only theoretical, but he did help me by buying from me one small piece of jewelry and directing me to a certain Mrs. Voranjinsky in Karlovy Vary.

So I backtracked one stop on the train and sought out this Mrs Voranjinsky at an address I had been given. She was apparently more experienced and up to date than Laricky. She was living with a small child and her mother. I saw no sign of a husband. She did not want money. She urged that it would be best, that I contact a Mr. Kurajsky in Cheb, and gave me several letters to carry to him. Thanks to this lady in Karlovy Vary I had a good meal and a rest. On her advice, I left behind my attaché case so that I would look less suspicious.

The following morning after breakfast I went back on the train to Cheb. I saw earlier how policemen in that border town examined everyone's documents and I prepared myself for an interrogation. However, the letters from that lady in Karlovy Vary apparently dealt with some sort of official matter, which Kurajsky was to take care of, and as a result once they saw the letters, they left me alone.

In Cheb I asked for directions and found this Mr. Kurajsky. He greeted me cordially, gave me some food and asked many questions about people I know in Prague. He was obviously testing me and when he was satisfied that I really knew people important to him, he seemed to relax. He promised to obtain a guide for me. After lunch he left and returned shortly with a strange, rough looking man, called Vincent, who took me to a small apartment. In the beginning he seemed wary of me but eventually warmed up and talked incessantly. He kept dropping names of people who he said were prominent in the movement and whom he knew personally. Finally he said he had to leave to visit some relatives and that I should stay there. He promised to return at night. I ate quickly in a nearby restaurant and bought a few groceries as there was nothing in the flat. I felt it best I not be seen often on the street.

Vincent returned and immediately began asking for money, and whether I have any gold that he could buy from me. I sold him two small gold bracelets for 1300 Crowns. Now I had about 10,000 Crowns all together. I stayed in that apartment overnight with an old man, who taught me how to play "66" (I already forgot how) and I went to sleep early. I felt well altogether, although it was cold. In the morning of March 28, I cooked and ate breakfast. Then I could do nothing but read and wait. Vincent came and wanted more gold. I gave him my cigarette case, which I think was only gilded.

Vincent said we would have a meal with one of his friends. We ate at the nearby restaurant, and later he informed me that we will be going that night. I packed a small case with my underwear and pajamas and dressed warmly with all the rest of my clothes. For about two hours we just sat and waited. I was very nervous. For a while we played cards and then we set out. We went to a small house where there was a man, who was Czech, and a woman who was German. They gave me supper, which was very good and nutritious. The woman apparently wanted to come along. They argued about this in an adjacent room. It was very unpleasant for me. At last they settled that we would go without her, which seemed to me a good decision. Around 8 PM, while it was still twilight but almost dark, we set out. We proceeded slowly, almost as if we were out for a stroll, walking through the town and into an estate that belonged to some German. He must have been quite wealthy for everything looked new. Vincent told me that the property was worth about 5 thousand Deutsche Marks, if he could find a buyer. Then he took me aside and as we sat hiding in some underbrush, he explained that Czech money is practically useless in Germany and that I should leave it all there. I gave him my money, saving only 1800 Crowns for myself. He claimed he would use it to pay his helpers, and he did not know how much they would want. If there was any left he would send it back to you, Pavlo, in Prague. I wrote a small note saying that if this reaches you, you should know that I am safe. I also wrote a note to the lady in Karlovy Vary asking that she send my attaché case to you. He promised to mail the notes that evening after he returned. I had some trepidation about making any trouble for you, even though I didn't sign my name. I knew you would recognize my handwriting.

At about 10:30 PM we finally left the estate and walked into a large field. It was very dark. While crossing a small mound I stepped with one foot into knee-deep water. Strangely I took little notice of it and don't even remember when it dried. I was really tense and nervous, constantly alert for the possibility of surprise everywhere. However, one cannot long sustain such a state of tension, and after a while as we kept walking briskly, I relaxed. Lights from Cheb twinkled behind us. I had no clear concept of where we were. We would walk for a while along some path and then through underbrush and then across a field. The ground was dry and we were able to walk freely. I have no recollection of being cold. Vincent carried two small suitcases and I my small bag.

After trekking through a plowed and particularly large field we came upon a paved road. We waited for a while before Vincent signaled for us to cross it. I suddenly realized that we were at a small airport and that this was a runway. We circled around two towers, which I supposed were guarded by soldiers. Vincent strode ahead with no apparent concern. He obviously knew the way. After a while we came to a fence and walked along it. It seemed to me we were moving back toward Cheb. I started to think that Vincent might be pulling some sort of dirty trick on me. Then we came to an underpass, just as the moon began to shine brightly. Vincent stopped and sat down. He told me to hide there and watch his suitcases, while he went to get something and would return. Before I had a chance to say anything he disappeared.

I was now alone and there was absolute silence around me. I sat scrunched under the small bridge so that I could not be seen from the road above. In the valley I could see lights from Cheb twinkling faintly. In the distance I heard a train whistle. I sat there about twenty minutes and kept thinking: Vincent got his money so what should he come back for? I tried to figure out what to do, and decided that if he doesn't return in an hour, I would walk along the fence back to Cheb. Then it occurred to me that I might examine Vincent's suitcases, and resolved to do it even if I would have to break their locks. The suitcases were heavy, but then that could have been anything even a load of dirt. Luckily they were not locked and were full of useful things: clothes, new shoes, identification papers and even some money. This satisfied me somewhat but still the waiting was excruciating.

Vincent returned after more than an hour, with a backpack. I felt reassured. We walked on with Cheb now to our right and it seemed that we were finally moving away from that town. Then we entered a wood. Vincent walked ahead. Several times I lost sight of him in the dark and had to whistle softly until we found each other. We climbed a steep hill and then again down into a valley meadow, which we crossed and then reentered the wood.

After about two and a half hours the woods started to thin and Vincent said that we might soon meet up with a friend of his who was supposed to reconnoiter the next section of our journey and tell us if there was any problem. We waited about thirty minutes. Vincent whinnied lightly like a horse. But there followed only silence.

So, finally we left the forest and entered a wide meadow. The moon was so bright it almost seemed like day. In a small valley to the left there was a house with a lighted window. We crossed one field, a road and then another. Strangely even though there was much moonlight I could not see Vincent but only sensed his presence because of a slight clanging of some metal object attached to his backpack. He was wearing a gray coat and hat that made him indistinguishable from the grayness around us.

We walked about twenty meters apart. In about an hour we entered a small village. It appeared to be completely dead and reminded me of that decimated village near Sojovice (14). We moved carefully along some remaining walls until we came upon a clearing with a large lake. The place was very damp. In the woods I slightly twisted my ankle and it now started to feel worse, but nothing could be done. I had to laboriously step over some broken electrical wiring laying on the ground, worrying about the possibility that some of it might still be live. After a while Vincent stopped pointed to a stone bench and said, "Why don't you rest here", and offered me a cigarette.

"I am not so tired that I can't continue and I would feel a lot better if we had the border crossing over with!" I said with some annoyance.

"Suit yourself," he answered, pocketing the cigarette pack, "but you are already in Bavaria."

Father's life as refugee

We had crossed the border in the middle of the village of Krusna Lipa. That town was probably evacuated because it straddled the border. Vincent was tired from carrying three bags, so we rested about twenty minutes and then we walked carefully through the remaining village streets to a large farm on the outskirts where Vincent knocked on the window. They opened the back door and led us down into their barn, which was a large space partly filled with straw, three horses and ten to twelve cows. I felt comfortable and warm, and very tired. We stretched out on the straw and Vincent immediately fell asleep. I, however, had a hard time falling asleep. I tried to assess the condition of my ankle and finally laid down on my side, trying to keep my leg immobile.

Toward the morning it grew chilly. At around 6 AM a woman from the house brought us breakfast, coffee and bread, which was very welcome. Vincent sorted out the items in his bags. He gave me 100 Deutsche Marks. I reminded him about the notes he promised to send to you and to Karlovy Vary and we parted. I then quietly slipped out and walked, as Vincent had instructed, to a nearby paved road where I saw the first German sign, "Nach Waldsam". I was to wait there for the authorities. It was about 7 AM. There was a town visible on the horizon.

An American Jeep arrived with a painted yellow stripe along its side. I waved to it and they called to me. There were four American soldiers. One of them spoke German. They took me into the Jeep in a casual way and asked me who I am. I told them that I had just escaped because I was in danger of arrest. We conversed for a while. They asked if I knew anything about a place called Jachym and whether there are any soldiers there. I told them that I thought that place was abandoned in 1945 and that no one was allowed in there. They then asked me if I, as a lawyer, approve the confiscation of German properties. When they drove very close to the border. I asked if they want me to cross back. They laughed at my worry. We finally went to a border station. Some German officials asked me to explain exactly where and with whom I crossed. I said I wasn't sure since the whole area was unknown to me, and my guide, whom I paid for the job, did not identify himself. The American police then took me to their headquarters, where I was assigned to some official, who spoke French. By then I was a bit confused with all the language barriers, but he was quite sympathetic, and I had occasion to see him often later. They took me to the CIC (Counter Intelligence Corps) headquarters, in the town of Selb, to register. He said I have to report to a refugee station called Hochunwort in Bavaria, and that an officer would drive me there.

Since I had endured police interrogations and jail, I thought nothing could surprise me, but Hochunwort was a shock. From a distance the large barracks did not seem so bad. Everything looked relatively clean, plenty of windows, smoke rising from chimneys, roads cleared. However, on closer inspection, it was horrible: nothing but one huge space; a wood floor on which were lined up crude mattresses stuffed with wood shavings; no pillows; two or three chairs for each group of about ten people; one table for the entire room. At least it was comfortably warm. I looked around seeking any potential friends. I finished some food that I still had with me and walked around trying to learn how things worked there. I was told that it was necessary to register to receive a blanket and a meal ticket, but that registration took place only in the morning. So if anyone arrived later, as I had, he had to spend the night on a bare bench in the dirt and cold. Luckily, a fellow Czech, named Jugoslav, offered me an unoccupied mattress next to him and lent me a blanket. I was very tired and around 6 PM I rolled up in the blanket, covered myself further with my coat and slept till the morning.

In the morning, after I signed in, I learned more about the misery there when I tasted the food. It consisted of some sort of stew, mostly potatoes and a piece of hard bread. We were given no eating utensils and had to use sticks or rolled up paper to help get the food in us.

Trying to understand the whole set-up, I talked to as many people I could and learned that we were in the custody of American police, CIC department, for processing escapees. We were waiting for proceedings in which they would evaluate evidence of our circumstances. There were also some representatives of the Bavarian Red Cross, who ironically had previously been assigned to take care of Germans on trial in Czechoslovakia. Germans, mostly those from the Sudetenland, comprised close to 80% of the personnel there, and it soon became clear to me that they felt about us Czechs the way most dogs feel about cats.

In the kitchen from which our meager rations emerged daily there was a large sign in German: "Any German woman who starts anything with a Czech will have her head shaved." Sure enough I soon spotted a woman, who had been friendly with some of the fellows, being pulled in the back, like a sheep to a shearing. She then came out with a kerchief wound around her head.

I heard many complaints. Some were unaware of the situation in Prague. They did not believe me that there was no official opposition to the Communists in Prague, not until some new people from Switzerland and others from Germany confirmed it.

At first it seemed that the Americans did not have any concerns about us and that the good reputation Americans had for caring for refugees was unfounded, but later I came to understand the difficulties of our situation: the whole post war operation in Europe was enormous and we Czechs were only a small drop in the sea of DP's (displaced persons) fleeing from different countries. There were about five hundred thousand DP's in Germany alone. We knew nothing about the International Refugee Organization (IRO) or any other help we could get. It was very difficult to get any information or money etc. We were free to leave the DP camp, but as it was I had to remain there until I could find some way to live. It was easy to feel helpless. I was very glad that we did not all leave together as a family. If I there were a city near and I had some Deutche marks, I would have been able to go to Frankfurt or maybe to Hamburg to try finding work. Thank God, I was able to get through that misery and save you from it.

I ran into some luck at the end of March when I met a certain Ms. Nesverbon. She seemed a peculiar character, always speaking flattery and at the same time eager to gossip, always saying that she never wished to harm anyone. Nevertheless I sensed that her basic motivations were good, so when she offered to take me to live with a German family and assured me they were very respectable people, I took her up on this. I was anxious to have a decent meal and to be able to sleep like a civilized person. The weather was getting colder and so were the barracks. They were always trying to save on fuel. I went with the Nesverbon woman and two others to town, where we stayed with a family named Parstim in a clean and pleasant place, and had good meals each evening and morning for two days. Then being determined to resolve my situation, I returned to the barracks. There I got to know some new people, who seemed particularly nice, though some talked too much. I thought about the saying: "Be careful what supposed friends really want".

One new friend said that he could exchange my crowns with someone in the nearby city of Cham. So I went there with him. It turned out to be a wild goose chase, but in Cham I ran into my border-crossing guide, Vincent, and asked him again if he sent my messages. I also asked if I could purchase from him some DM's. He promised to try to get some. But I felt he would not. There is a beautiful hotel in Cham, so I went there asking for some sort of position. I told them that I am expecting to be joined soon by my family. This effort actually calmed me, even though I had to smile at its futility. I returned to the Parstim family and stayed for another day. In retrospect I must admit that having their help was what kept me going.

Eventually, I made some headway with the Americans, and they introduced me to a certain Mr. Klimeshin. He seemed to me some sort of smuggler or secret agent, since he regularly traveled to Prague, but I was grateful as he agreed to post a letter for me in Prague, in which I wrote to you about my safety.

I was visiting the Parstims again on the second of April, when another member of the family showed up and I found out one reason why they were being so nice to me: I had defended him in Prague during some small trial. We had a good dinner during which I learned that Mrs. Parstim spoke a bit English and for that reason she had obtained a position in a hospital. Unfortunately my ability to speak German and French did not help me get a job.

One day my friends and I were moved from the communal barracks into a little hut, number 37, that stood by itself in a field. There were seven of us, five men two of whom had their wives with them. We were told we would be transported to a camp in Schwabach, a city near Nuremberg, which was reported to be more stable. We were uncomfortable and hungry, having only bits to eat. Once they gave us salami that had gone bad. I felt weak and tired. I don't know the cause but I started to feel sick. Luckily it subsided in two days, and I was able to venture outside with some of the other men. We thought we might gather wood to burn for warmth in an iron stove there, but we couldn't find even a small twig. One found a wooden bench, but no one could break it up for kindling. People started to complain and say crazy things.

After two days, as more of us were starting to lose hope, an order came to get ready for transport that afternoon. We were supposed to get something to eat before our journey, but apparently someone stole that food, for we received nothing. At 2 PM we were loaded into completely bare railway cars. So we had to stand, sit or lie down on the floor. There was no heat. In the day it was tolerable but we knew it would get chilly toward evening. Still we were happy to be going somewhere at last. We knew Schwabach was about 170 kilometers away, a journey of about five hours, and so we figured we would arrive at about 8 PM. But although the train started as soon as we were loaded, it first backtracked to the town of Hof. There we stood for about six hours before it started again. The train rolled along briskly, and it started to get really cold. We had left the large door partially open to get fresh air and some light. But now we were forced to close it. The atmosphere in the interior became ominously dark and dank. At last in some station, we were allowed out to buy coffee. It was bitter but most importantly hot. I and some others also managed to get some soup.

I want to relate how three of us came to share a certain treasure. One time two of my roommates found a rusty spoon on a pile of garbage. We scrubbed it with sand and managed to make it shine. To eat with a spoon was really something. It was a privilege for me to be allowed to carry it. I am sorry that I somehow lost it since then. It would have made a good souvenir.

We were loaded back, and the train went on and on. It was now eight PM and totally dark. Some huddled together for warmth. It was hard for me to sit. So I paced inside that wagon most of the night. Sometime around midnight we reached Nuremberg, and stood there for about 2 hours. Feeling wretched, I thanked God once again that you and the children were not with me. How much the little ones would have suffered. At last we started moving again and after about three more hours we reached the refugee camp of Schwabach. There we were marched on foot to the camp headquarters.

We were greeted by two rather officious persons, including an unpleasant, old soldier. They delivered rules and made us fill out a questionnaire designed to establish evidence of our status. Still it was wonderful to be in a dining area where it was pleasantly warm. By the time they were done telling us what was allowed and what was forbidden, it was morning. We were now in Austria under the auspices of the Red Cross.

We were sent to a barrack number 19. There were at least twelve people per room, and protests against us started immediately. One common pair loudly cursed the official for assigning more people where there wasn't a single free bed and not even an extra blanket etc. We wound up in room C, where conditions were essentially the same as in Hocheworf -just bare mattresses. Eventually, though, each of us received two blankets and even a bowl for food. One spoon was provided for several people to share. In the center space was an iron stove for which each of us got small rations of coal. There was dirt everywhere. I succeeded to grab a spot by a window, where I managed to arrange my few clothes and papers, and tried to start a regular routine. I wrote to everyone I knew, hoping to find some route out of my present status. I tried to learn about the conditions in this camp and its surroundings. Knowing that I must earn some money, I kept trying to figure out how I could show myself able to rise above the pitiful and often nasty state around me. At that point I didn't fully realize what a miserable state I myself was in.

We were told that we must apply for a work permit, for without it we would not be allowed to leave the camp. (This was not really so, for I had already left the camp several times.) We were all photographed for this card and when I saw my picture I got a shock. (We had no access to a mirror, and I had been shaving using my indistinct reflection in a window.) Now I saw an ugly, skinny, old man. My state was also evident from my clothes. My shirt hung on me so that it reminded me of the time when little Dadunka (11) put on my shirt. I had to make new holes to shorten my belt.

A certain petty officer sometimes talked to us escapees. He was very polite and asked me what languages I spoke. One day he told me that he knew a way how I could earn some money and drove me to people he knew for this purpose, but nothing came of this.

On our return to the camp, I was amazed when he picked up my old guide, Vincent. I immediately asked him if he sent the notes I had requested. He nodded but I doubted if I could really trust him. In the car I tried to make conversation. I asked both of them about what they do there and about their families. Being friendly sometimes has a magical effect and they decided to take me to their home. Then they helped me to sell the little ring with a small diamond, which I had kept hidden in a breast pocket. The deal went through a business that was very much like the one that used to be on Jungmanova street [i.e. a kind of pawn shop]. Such dealers manage somehow to stay in business even during the awful times we were in.

Now that I had a bit money (I got six thousand marks for that ring), I went to see an old friend, named Jansyn, who was then living in Nuremberg. Jansyn said he had met an American general named Leroy Watson and that this general had helped some Czechs before. So we went to meet General Watson, whose address we obtained at the American headquarters. It was a long walk to the outside of what was left of the city, but our trek to his house paid off. A lady there greeted us warmly, gave us good coffee and warm cheese kolace, sandwiches and even cigarettes. It was a treat that I did not have for months and so of course we ate everything eagerly. It seemed that this lady was used to this kind of reaction.

This contact, and people I met subsequently through a movement that worked to help displaced Czechs, connected me with a company producing a Czech language newspaper in Frankfurt. This newspaper, New Yorksky Listy, was published in New York City, for the Czech immigrants there. And that's how I escaped having to live in the DP camp, and even received some perks that employees of American enterprises received in the American Zone of Occupied Germany.

With money father saved from his salary at that Newspaper, he managed to fund an underground escape for my mother, sister and me. It had to be a clandestine trip on foot because the Communist regime would not allow us to leave openly, despite my mother's many efforts. Eventually, we all immigrated to the USA and became naturalized citizens, where father and mother struggled in various jobs to finance education for my sister and me.

Looking back at the state of Germany after the war father wrote:

In Nuremberg, I saw for the first time, what total destruction war can cause. In comparison, the devastation in our part of Prague during the bombing at the end of the war seemed like child's play. I was familiar with the old Nuremberg, and now that city simply did not exist. The American bombers had wiped it totally away just like cleaning a writing slate. It reminded me of a bees nest broken into by a bear bent on totally emptying it of honey. There were only piles and piles of rubble, remnants of walls, dust and misery. Here and there someone had tried building a provisional shelter. The worst to me were large mountains of ruins, apparently never touched, under which undoubtedly lay many, now all hopefully dead, bodies.

Hans Sachs Plats (16) was completely demolished. Around it not one house was recognizable. Amidst the pile of what I thought must have been the large hotel there, a statue of Hans Sachs, somehow spared, stood holding a bronze pen in its hand. The sight brought me to tears. Surely any German seeing such a desert would feel resentment. The Nazi leaders must have realized early in 1945 that all hope of victory was gone. Why did they not surrender and at least spare some people and things? But Hitler and the whole Fascist idiocy are not the only objects of my curses. I also blame the many ordinary people who saw some good in the movement, and thought that the clique around Hitler meant well after all.

The relationship between Germans and Americans after the War was peculiar. They saw each other's customs as strange, which led to a kind of mutual derision. Many American soldiers appeared to view German culture as old and stuffy, while many Germans considered Americans to be juvenile. Still the German people had a certain slavish respect toward Americans, because they knew their immediate future depended on their financial help. Actually, the one thought that preoccupied most Germans was the fact that they were living like dogs: There were shortages of all good things. All that seemed around was shriveled potatoes, wilted vegetables and terrible watery beer.

References and Footnotes

1. My mother's given name was Pavla. In Czech when a name is used to address someone it is put into the "accusative case". So in this instance Pavla becomes Pavlo.

2. Julian Duris and Vaclav Kopecky were prominent in the fight for the Czechoslovak nation against outside usurpers.

3. The history of the Schwarzenberg family dates back to the middle ages. See House of Schwarzenberg, Wikipedia. (The Soviet Communists had no respect for our history.)

4. Jan Masaryk was the son of Tomas Masaryk, the first elected president of Czechoslovakia. Eduard Benes was our second elected President

5. Josef Sveik is a fictional character invented by Jaroslav Hasek for his comic satire, "The Adventures of Good Soldier Sveik".

6. I imagine father as lawyer knew about confiscations of private property (like what happened to our telephone line, later). I myself saw that the Soviet flag and pictures of Stalin became prominently displayed in schools and in store windows.

7. President Benes was succeeded by a Soviet puppet, Klement Gottwalt.

There is evidence that both Jan Masaryk and Eduard Benes were murdered by the insurgents. Corroboration for their murders can be found in Madeline Albright's book, *Madame Secretary,* published in 2003.

8. Pankrac is a large prison in Prague.

9. Politically motivated executions were increasing, which I am sure father knew about because of his old friends at the courts where he had spent years as a lawyer.

10. Mr. Vana, who I think was some official in the railroad that crossed the border between Czechoslovakia and Germany, later delivered clandestine letters to mother, which assured her that father was alive and safe. In these letters, father used a previously agreed on code signature. Mother destroyed each letter immediately after she read it.

11. Sarka was my sister's given Czech name (in the US she changed it to Shari), and Dadla, or sometimes Dadunka were my childhood nicknames.

12. Cheb is a city at the border between Czechoslovakia and Germany.

13. Sokolov is a town in the Karlovy Vary Region of the Czech Republic.

14. Sojovice is a village near Prague where we often went for summer vacations. I don't know what "devastated village" father is referring to at this point of his narrative, but it may have been Lidice, the town destroyed by Nazis in 1942 as a punishment and warning for the assassination of a prominent Nazi.

15. Jachym is a small site near the Czech-German border, which is dedicated to Saint Jachym.

16. Hans Sachs Plats (i.e. square) is named after a German poet and playwright from the sixteenth century.

www.ingramcontent.com/pod-product-compliance
Lightning Source LLC
Chambersburg PA
CBHW020921140626
46545CB00015B/1108